Contents

Light and sound

Light and sound both play a huge part in our lives. They allow us to see the world around us, to hear what is going on, and to enjoy sights and sounds, such as paintings and music. Most animals have sensitive eyes and ears to detect light and sound. These are just as important to animals as humans, because animals use them for hunting and for sensing danger.

Light and sound are both forms of **energy**, and can be changed into other sorts of energy, such as **electrical energy** and **mechanical energy**. Light and sound have another thing in common – you can think of both as **waves** which spread out from where they are made.

Light sources

Light is made by a change of energy. For example, in a fire, light is made from the **chemical energy** in the material which burns, and in a light-bulb, light is made from electrical energy. Light travels away from its source in straight lines called rays. You cannot see these rays unless they enter your eyes, but you can sometimes see rays of sunlight as they illuminate dust in the air.

During the day, most of the light we see comes from the Sun. The amount of light that reaches the Earth from the Sun is enormous, but it is only a tiny fraction of the light that the Sun gives out. The light is created from **nuclear energy** in the centre of the Sun. Its energy supports most of the Earth's life and provides the energy that makes the world's weather happen.

Catching sunlight

Solar energy is energy captured from sunlight. Solar cells turn light directly into electricity. Solar panels turn light into heat energy and use it to heat water. The amount of energy even in bright sunlight is quite small, so large panels are needed to capture a useful quantity of energy. For example, a panel of solar cells 1 metre by 1 metre is needed to produce enough electricity to work a light-bulb ... and only if the Sun is shining brightly! However, solar energy is very environmentally friendly.

Blocking the light

A **shadow** is made when an object blocks the path of light rays, forming a dark area on the side away from the light. If the unhindered light then hits a surface, you can see the dark outline of the object.

The Earth itself forms a large shadow because it blocks sunlight. Each night, the part of the Earth where you live moves through the shadow. During the day, shadows made by sunlight move as the Earth spins and the Sun appears to cross the sky.

◀ Long shadows form on the landscape in the early morning and late evening, when the sun is low in the sky. At midday in summer, the shadows will be right under the trees.

The speed of light

Light travels fast ... very fast. In a **vacuum** it travels 300,000 kilometres every second. Light coming from close-by objects gets to you almost in an instant, so you see things as they happen. But even at such high speed, light from very distant objects can take a long time to arrive. For example, light from the Sun takes eight minutes to reach the Earth. This means that where the Sun appears in the sky is actually where it was eight minutes ago! The speed of light is a very important number in **astrophysics**, and according to Albert **Einstein's** theory of relativity, nothing can travel faster than the speed of light.

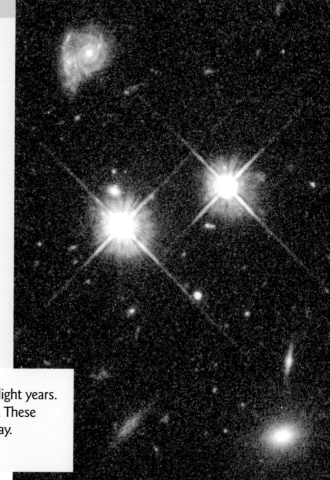

▶ Astronomers measure the vast distances in space in light years. One light year is about 9.5 million million kilometres. These **quasars** are more than 13,000 million light years away.

Light for seeing

We see things because light rays coming from them enter our eyes. At the back of the eye is a layer of special cells called the retina which detects the light and sends signals to the brain. The picture you see is built inside your brain. At the front of the eye is a **lens** which organizes the light to make a small picture of the scene in front of the eye on the retina. You can find out how lenses work on page 10.

You can find out how lenses work on page 10.

SCIENCE ESSENTIALS

We see objects when the light rays which they emit or reflect enter our eyes.
When light rays hit an object, they are either transmitted, absorbed or reflected.

Light and materials

We see sources of light, such as light-bulbs, because light rays from them go straight into our eyes. We see objects which are not sources of light because light rays from other sources hit them and then go into our eyes. A simple example of this difference can be seen in the night sky. We can see stars because they make light, and planets and moons because light rays from the Sun hit them, and some of the rays bounce off towards us.

When a light ray hits an object, it interacts with the object in one of three different ways. It either bounces off the object, which is called **reflection**, or it disappears into the object, which is called **absorption**, or it goes through the object, which is called **transmission**. For example, white paper reflects the light that hits it, plain glass transmits the light that hits it, and black paper absorbs the light that hits it.

Materials that do not allow any light rays to pass through them, such as wood, are called **opaque** materials. Materials that let all light rays pass through, such as glass, are called **transparent** materials. Materials that let some light rays through but scatter them about, such as tracing paper, are called **translucent** materials.

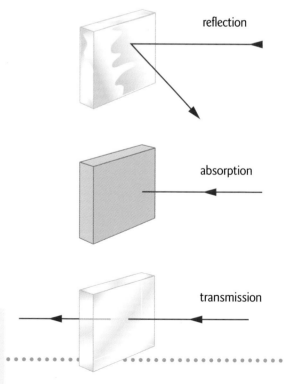

reflection

absorption

transmission

► Examples of light being reflected (top), absorbed (middle) and transmitted (bottom) by different materials.

Animal lights

In the depths of the ocean there is a vast, dark world that little light can reach. The many creatures that live there produce their own light, called **bioluminescence**. Some make light by **chemical reactions** inside special light organs in their bodies. Others have light organs that contain glowing bacteria. They use their light to send signals, either to find a mate, frighten off an attacker, attract their prey, or camouflage themselves.

Some creatures on land also glow with their own light. The best known are glow-worms (fireflies), which are a type of beetle. They flash their lights in the darkness to attract a mate. Each species has its own pattern of flashes so that members of the same species can recognize each other's signals.

▼

Many deep-sea fish produce their own light to help them survive in the depths of the sea where no natural light can reach. This deep-sea viperfish has rows of special luminous structures along the length of its body. These are called photophores.

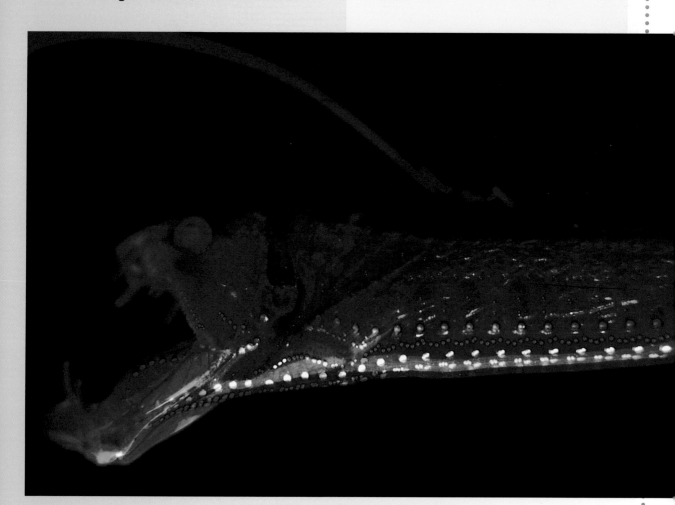

Reflection and mirrors

Most objects are visible to us because light reflects off them and into our eyes. Different objects look different because the materials they are made of reflect light in different ways. The shape and texture of the object matter, too.

SCIENCE ESSENTIALS

Light reflects well from shiny surfaces.
Light is scattered by dull surfaces.
A mirror is a perfect reflector.

Shiny and dull

Imagine lots of light rays coming from a small light-bulb, all close together and parallel to each other. When they hit a perfectly flat surface, such as a piece of polished metal, all the rays bounce off in the same direction, and stay parallel. If these rays enter your eye, they look as though they have come from a source of light in the metal, so the metal looks shiny. Now imagine the same rays hitting a surface which is not perfectly flat, such as a piece of tissue paper. The rays still bounce off, but this time they are scattered in different directions. The rays are no longer parallel, so the surface looks dull. This is called diffuse **reflection**.

Mirrors

A **mirror** is a perfect reflector. It reflects all the light that hits it and keeps the pattern of rays the same. This means that rays of light coming from an object that hit the mirror are reflected and continue going as though they had come from an object behind the mirror. In fact, this makes the object appear to be behind the mirror. The picture of the real object that you see in a mirror is called its **image**. The only difference between the object and its image is that left and right appear to be reversed.

Most mirrors are made of a sheet of glass with silver paint on the back. The light reflects off the silver paint which has a very flat surface because the glass is extremely smooth.

Curved mirrors

A mirror with a curved surface makes objects look larger or smaller than they really are. A **concave** mirror has a surface which curves inwards. It reflects rays towards each other (it makes them converge). A **convex** mirror has a surface which curves outwards. It reflects rays away from each other (it makes them diverge). Curved mirrors are used widely in optical instruments such as telescopes.

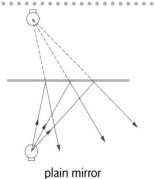
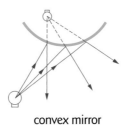

plain mirror   convex mirror   concave mirror

▲ In a plain mirror, objects seem to be behind the mirror. They seem the same in a convex mirror, but appear smaller. In a concave mirror, they seem to be in front of the mirror.

Super-accurate mirrors

If you look closely in a mirror, you will be able to see faint 'ghost' images a few millimetres away from the normal image. These occur because some light reflects from the front face of the glass. This does not really matter in the bathroom, but in a reflecting **telescope** mirror it makes extra images of stars, which can be confusing. Reflecting telescopes have a large concave mirror, made from glass. To solve the 'ghost' image problem the front face is covered with a super-thin coating of metal, such as aluminium, which forms the reflecting surface.

The larger a telescope mirror is the better, because it collects more light and gives a brighter image of the stars. The largest ever made is six metres across. Telescope mirrors are made from special glass to stop them expanding and contracting as the temperature changes. Making these huge, curved mirrors accurately is not easy because they become very heavy. One answer to this problem is to use several small telescopes instead of one large one, and then combine the images from them on a computer.

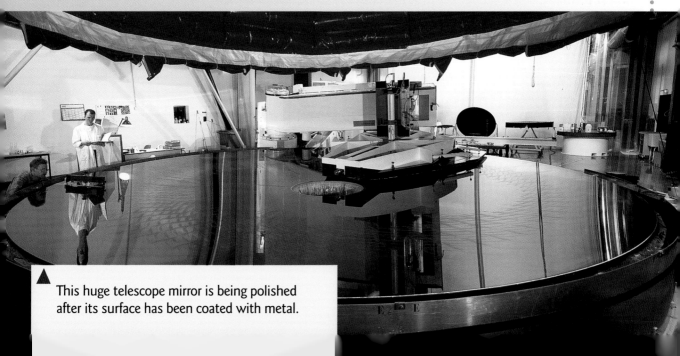

▲ This huge telescope mirror is being polished after its surface has been coated with metal.

Bending light

When a light ray crosses the **boundary** from one **transparent** substance to another, its direction changes. This effect is called **refraction**. For example, when a light ray from an underwater object comes up through the surface and into the air, it gets bent. The only time there is no refraction is if the light ray hits the boundary at right angles.

SCIENCE ESSENTIALS

Light rays change direction when they cross the boundary between one substance and another. This is called refraction.
A lens bends light in an organized way.

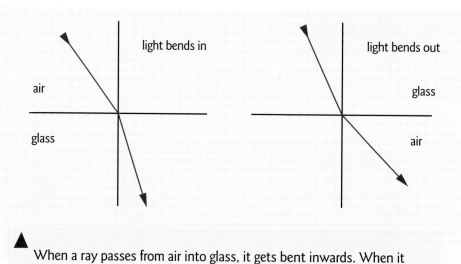

light bends in

air

glass

light bends out

glass

air

▲ When a ray passes from air into glass, it gets bent inwards. When it passes out of the glass into the air, it gets bent outwards again.

Apparent depth

Refraction can make an object appear to be in a different place to where it really is. For example, when you look down into water, objects under the water appear to be higher up than they really are. This is because the light rays from the objects are bent towards you as they leave the water.

Lenses

A **lens** is a specially-shaped piece of glass or plastic that bends light in an organized way. The type of lens you are most likely to see is called a **convex** lens. It is the type of lens found in a normal magnifying glass. It has surfaces that curve outwards, so the lens is thicker in the middle than at the edges. A **concave** lens is the opposite shape. It has surfaces that curve inwards, so the lens is thinner in the middle than at the edges.

Making images

A convex lens bends light rays towards each other, and so is often called a converging lens. Two parallel light rays passing into one side of the lens are bent so that they meet on the other side at a place called the **focal point**.

A concave lens bends rays of light away from each other, and so is often called a diverging lens. Two parallel light rays going into one side of the lens are bent so that they appear to come from the focal point on that side.

Total internal reflection

Imagine a ray of light travelling inside a block of glass. Let's look at what happens when it reaches the surface of the glass.
• If the ray hits the surface straight on at right angles, it goes straight through, without changing direction.
• If the ray hits at a slight angle, it goes through the surface but gets refracted.
• If the ray hits at a large angle, it does not go through at all, and is reflected – the surface of the glass acts like a mirror. This effect is called total internal **reflection**.

Total internal reflection is very useful because it does not create ghostly double images like a normal mirror does. Optical instruments such as binoculars and **periscopes** use **prisms** instead of mirrors to turn light rays around corners. Cycle reflectors and cats' eyes reflect light by total internal reflection too.

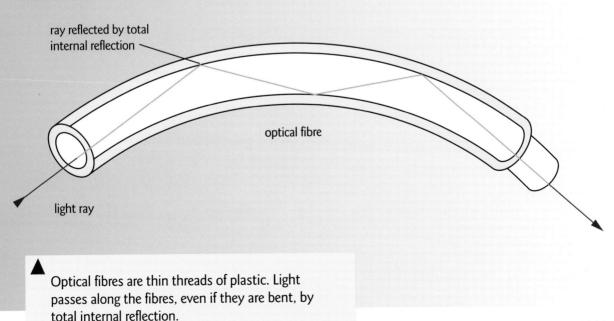

ray reflected by total internal reflection

optical fibre

light ray

▲ Optical fibres are thin threads of plastic. Light passes along the fibres, even if they are bent, by total internal reflection.

Optical instruments

Optical instruments are devices that make use of **mirrors** and **lenses** to make **images** of objects. Examples of optical instruments are cameras, **telescopes** and **microscopes**. Some optical instruments are simply used to capture light. Others enable us to see objects that we cannot see with the naked eye.

Capturing an image

The job of a camera is to record a scene. To achieve this it uses a **lens** to create an image on film of the scene the camera is pointing at. The lens collects light rays from the scene and bends them to make an image inside the camera. The lens in the front of your eye makes an image on the retina at the back of your eye in the same way.

When the camera focuses, either automatically or manually, the distance between the lens and the film is adjusted so that the image falls accurately on the surface of the film in the back of the camera. The film is able to record the image because it contains chemicals that change when the light of the image hits them. When the film is processed, the image appears.

Seeing the small

A microscope allows you to see very tiny objects in detail. You cannot see the details with the naked eye because the lens in your eye cannot bend the light enough to focus an image onto your retina. A magnifying glass is the simplest type of microscope. Its **convex** lens bends light inwards, allowing you to move your eye closer to the object you are examining. Most microscopes have two lenses. The first lens, called the objective lens, makes an image inside the microscope. The second lens, called the eye lens, magnifies that image.

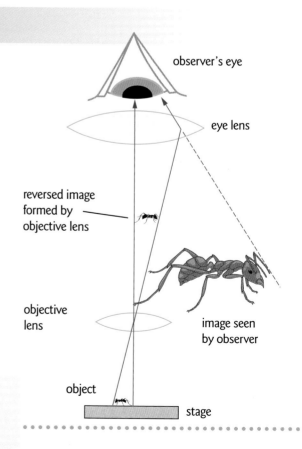

observer's eye

eye lens

reversed image formed by objective lens

objective lens

image seen by observer

object

stage

▶ A microscope bends light rays coming from an object so the object appears to be much bigger than it is.

12

Seeing the distant

Telescopes make distant objects look larger. The simplest type of telescope is the refracting telescope, which has two lenses. The first lens, called the objective lens, makes an image of the distant object. The second lens, called the eye lens, magnifies that image. A reflecting telescope has a large **concave mirror** instead of an objective lens.

Digital images

Digital cameras take photographs electronically. Instead of film in the back of the camera, there is a special **microchip** called a charge-coupled device (CCD) which records the image. The face of the CCD is covered by a grid of thousands of tiny squares called pixels. Each square is charged with electricity before the image made by the camera is allowed to fall on the grid. The colour and brightness of the light of the image affects the charge in the pixels. Electronic circuits detect the remaining charge and so record the colour and brightness of the image at each pixel. This information can be loaded into a computer, where the images can be viewed, manipulated and printed.

The CCD in an average compact camera has a grid a few hundred pixels wide, with perhaps a few hundred thousand pixels in all. Telescopes and microscopes use CCDs too. They are much larger and more detailed than those in cameras, with a total of tens of millions of pixels. The images from these telescopes and microscopes can be analysed and **enhanced** by computers.

▼
This close-up photograph shows a charge-coupled device (CCD) from a digital camera. The central, flat section detects the light.

Colour

The light that comes from light-bulbs or the Sun looks white, and is called white light. However, white light does not really exist. In fact, it is a mixture of equal amounts of many different colours of light which combine together to look white.

Rainbow colours

Sometimes white light gets split into its different colours, making the colours visible. This happens because the different colours are refracted by slightly different amounts when they cross a **boundary**. For example, when white light crosses a boundary, the red part is bent more than the blue part. This effect is called **dispersion**. This is what happens when a rainbow forms.

Sunlight is dispersed as it is refracted around rain drops.

The colours in a rainbow are called the colours of the spectrum. They are red, orange, yellow, green, blue, indigo, violet. The colours do not change abruptly from one to the next, but blend together smoothly.

How objects have colour

You already know that when light hits an object it is either reflected, transmitted or absorbed. But why does one object look different to another? The answer is that the objects affect the light that hits them in different ways. For example, a green apple looks green because it reflects just green light. It absorbs all the other colours of light. This means that only green light is visible when you look at the apple.

► When white light hits this apple, all the colours of the spectrum are absorbed except green, which is reflected. This makes the apple look green.

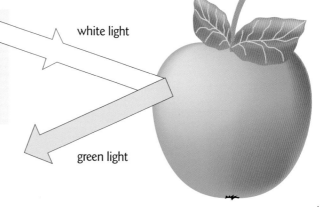

white light

green light

green apple

Filters

A **filter** is a piece of glass or plastic which only lets some colours of light through. It absorbs all the other colours. For example, a red filter only lets red light pass through, and stops all the other colours. So when you shine a white light through a red filter, only red light comes out on the other side.

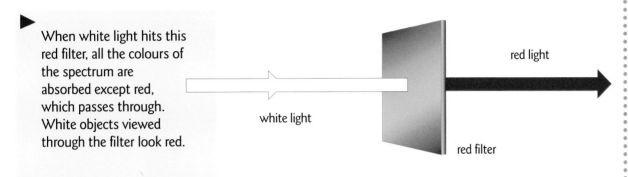

► When white light hits this red filter, all the colours of the spectrum are absorbed except red, which passes through. White objects viewed through the filter look red.

white light

red light

red filter

Flames and spectroscopy

When any substance is heated to a very high temperature, it begins to emit light. The colour of the light can help to determine what the substance is. One simple test in chemical analysis is the flame test, which is used to find what sort of metal there is in a substance. For example, a substance that contains copper burns with a green-blue flame, and a substance that contains sodium burns with an orange flame.

In fact, substances give out a mixture of different colours of light when they are heated, and in a flame test these are all mixed together. To identify the colours a device called a spectroscope is used. This disperses the light from the substance, showing which colours of the whole spectrum are present in the light, and in what proportions. This is called an **emission spectrum**. Every chemical **element** has its own characteristic emission spectrum. This means that a spectroscope can be used to find out what elements there are in a mixture.

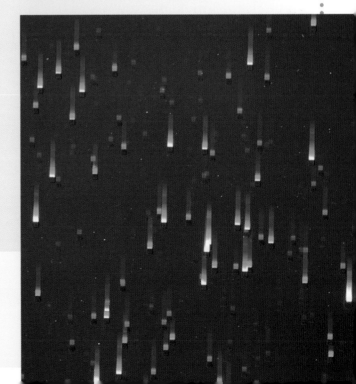

► Each tiny spectrum is made by splitting light coming from an object in space. Different objects (stars, quasars, and so on) create a different spectrum.

Primary colours

Although there is a range of colours in the **colour spectrum**, there are some special colours which can be combined together to create any colour. These colours are called primary colours. There are three primary colours of light.

SCIENCE ESSENTIALS

The **primary colours** of light are red, green and blue.
Any colours can be made by adding the primary colours in different proportions.

Primary colours of light

The primary colours of light are red, green and blue. By mixing these colours in different proportions, any colour of the spectrum can be made. This process is called colour addition because by adding different colours of light together new colours are made. Mixing red, green and blue light in equal proportions makes **white light**. An example of colour addition is in television. The screen gives out just red, green and blue light, but it adjusts the brightness of each one on different parts of the screen to create the coloured picture.

Colours of paint

The colours in inks and paints are created by **pigments**. You see these colours because of **reflection**. When white light hits an object, the pigments absorb some colours and reflect others. This is called colour subtraction. The primary colours of pigments are yellow, cyan and magenta. When they are all mixed together, all the colours of the spectrum are absorbed and you see black.

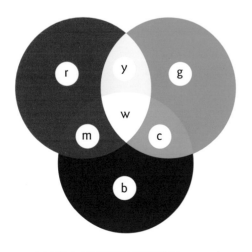

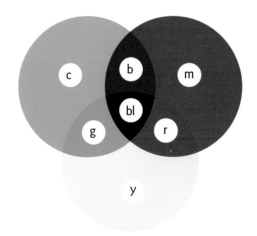

▲ Adding the primary colours of light (left) and pigments (right).

r = red; g = green; b = blue
c = cyan; m = magenta; y = yellow
bl = black; w = white

Colours in coloured light

In white light, coloured objects look the colour they are because they reflect certain parts of the white light. If they are illuminated by coloured light, they look a different colour. For example, a green apple illuminated by a red light looks black because there is no green light to be reflected and the red light is absorbed.

Printing in colour

If you look at the colour photographs in this book with a magnifying glass, you will see hundreds of tiny dots of colour. But there are only four colours – cyan, magenta, yellow and black, or CMYK for short. The different colours in the pictures are produced by adjusting the size of the dots. From a distance, the dots blend together to appear as colours. Black ink is needed because the other three make brown, not black, when they are blended together. Black is also normally used for the text.

When the page is printed, the four colours are added one by one, using the four coloured inks. Gradually the pictures are built up. A book like this is printed on a four-colour printing press. This printing process prints colour books and magazines with just four colours of ink. Colour ink-jet printers work in the same way by firing tiny blobs of coloured ink onto the paper.

However, CMYK printing cannot reproduce certain colours, so sometimes special colours are added to the basic four. This is often the case for printing fine-art prints, which need a subtle range of colours. Metallic colours, such as gold and silver, or **fluorescent** colours, are added with separate inks too.

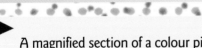

▶ A magnified section of a colour picture from an ink-jet printer. You can see the tiny dots of cyan, magenta, yellow and black ink, which are normally too small to see.

Using colour

Try spending a day looking out for colour – look at the decorations in your home, your clothes, colours of plants as you travel to school, signs around your school, books and the pictures and lettering in them. Try to imagine why all the objects you see are the colours they are.

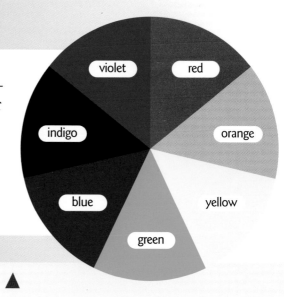

A sample colour wheel using just seven colours.

Colours for decoration

Interior decorations, clothes, books and magazines all use colour to look good. The trick of creating a pleasing colour scheme is choosing the right colour combinations – get it wrong and the result is almost impossible to look at!

If you think of grabbing the two ends of the **colour spectrum** and pulling them together so that red is next to violet, you get a colour wheel. Mixing colours that are opposite to each other on the wheel, such as red and green or blue and orange, makes for a contrasting, vibrant colour scheme.

Mixing colours that are close to each other on the wheel, such as blue and indigo, creates a harmonious colour scheme.

As you move along the spectrum from red to violet, the colours change from 'warm' to 'cool'. Decorating with warm colours, such as red and orange, makes a room feel cosy. Decorating with cool colours, such as greens and blues, makes a room feel cool and airy.

Getting the message

Colours are also good at getting messages across quickly. Flashes of colour in a bland background catch the eye and attract people's attention. Signs are more likely to be read if they are in red type or have a red border around them. Certain colours have come to have special meanings. For example, red is often used for danger, yellow is for warnings, and green normally stands for 'okay'. You can see these colours at work in traffic signals.

Bright red is a colour that is not often seen in nature or used in decorating, so a bright red sign stands out from the background.

Colours in nature

Animals and plants use colour to their advantage too. Many flowering plants have brightly coloured petals which advertise to birds and insects that there is nectar inside the flowers. As the animals visit flowers, they carry pollen from one to the other, helping the plant to reproduce.

In many species of bird, the male has brightly coloured feathers or other body parts for display during courtship, in order to attract a mate. Some animals use colour for camouflage, so that they can hide from predators amongst leaves or grass. Other animals, such as ladybirds, snakes and frogs, do quite the opposite – they have bright spots or stripes of red or yellow on their skins to warn predators that they are poisonous or vile-tasting.

Perhaps most amazing are those animals that can change colour at will. Lizard-like chameleons change the colour of their skins to match their background, so they can camouflage themselves almost anywhere. Cuttlefish have special droplets of different coloured pigments in their skin. By changing the sizes of the droplets, they change their colours. Cuttlefish are thought to send messages to each other by making ripples of colour move along their bodies.

▼
The bright colours of these poison-arrow frogs warn predators that there are chemicals in their skin which are deadly poisons.

Invisible light

Light is a type of **electromagnetic radiation**. This means that it is made up of changing electric and magnetic fields. Physicists often think of this radiation as travelling in **waves**, just as waves travel across water, spreading out from where they are made. Light is not the only type of electromagnetic radiation. It is just one member of a whole family of electromagnetic waves, called the **electromagnetic spectrum**.

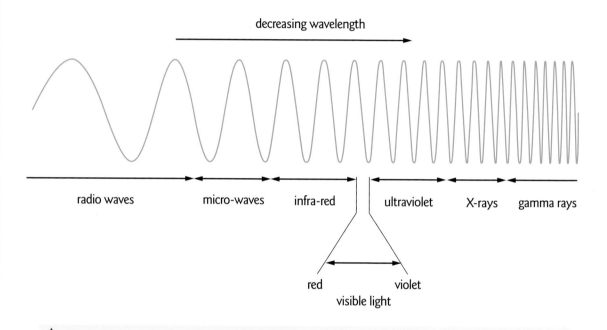

decreasing wavelength

radio waves   micro-waves   infra-red   ultraviolet   X-rays   gamma rays

red   violet
visible light

A simplified diagram of the electromagnetic spectrum. In reality the wavelengths at the left end are millions of millions of times greater than those at the right end.

The electromagnetic spectrum includes waves that seem very different to light, such as radio waves, microwaves and X-rays. But all these waves travel at the speed of light and carry **energy** away from where they are made. As you move from one end of the electromagnetic spectrum to the other, the **wavelength** of the waves decreases and the **frequency** increases (you can find out more about wavelengths and frequencies on page 24).

The part of the **colour spectrum** that makes up the light we see is called the visible spectrum. Each colour is made up of electromagnetic waves of a different wavelength. We see the different colours because our eyes can detect the different wavelengths.

Infra-red and ultra-violet

Sunlight contains two more types of radiation which are often included in the colour spectrum, even though they are invisible. Below red light is **infra-red radiation**, and above violet is **ultra-violet radiation**.

Infra-red radiation is a form of heat energy. It is given off by all objects, but gets stronger as the object gets hotter. You can feel infra-red radiation as warmth on your skin.

Ultra-violet (UV) radiation is the type of radiation that tans fair skin in the Sun. Small amounts of UV are good for you, but large amounts can cause sunburn and even skin cancer. Some security markings are made with special ink which only shows up when UV light is shone on it. If items that are marked with the ink are stolen and then recovered, the police can use a UV lamp to identify the goods.

The invisible sky

Taking photographs of space can tell us a certain amount about the stars, galaxies and other objects in our Universe, such as their positions. However, much more information can be found by collecting the non-visible types of radiation coming from space, such as radio waves and X-rays. Visible light coming from distant stars is often blocked by huge clouds of dust and gas far out in space. But other types of radiation from the stars can pass through the clouds, so looking for these can make the stars and galaxies 'visible'.

Infra-red astronomy shows the heat coming from objects in space. The most intense heat comes from the centre of our **galaxy**, where a great mass of stars are being formed. X-ray astronomy shows that X-rays are coming from all over the sky. Most of them come from hot gas in our galaxy. Radio astronomy shows up many things, including sweeping beams of radio waves coming from tiny dead stars called pulsars.

▶ Infra-red radiation is also used to photograph the Earth from space. This image shows the heat coming from part of Japan. Plants look red, buildings blue and water black.

Waves and sound

A **wave** happens when a movement is repeated again and again. The waves we see most often are the waves that move across the surface of an ocean or lake. However, light and sound move as waves too, even though we cannot see the waves in action. All waves carry **energy** away from where they are made.

Moving in waves

As an ocean wave travels past a point in the water, the particles of the water move up and down. As they move, they make the particles next to them move too, and so the wave moves through the water, even though the particles themselves do not actually move along with the wave. Because the water particles in waves actually move at right angles to the direction of the waves, water waves are called transverse waves.

The other type of wave is called a longitudinal wave. In a longitudinal wave, the particles move backwards and forwards in the same direction as the waves are travelling. An example of a longitudinal wave is a wave going down a long, slinky spring. Pulling briefly on one end of the spring pulls the coils at that end apart. Each coil then moves one way and then the other, as the wave moves along the spring.

Wave words

The amplitude of a wave is the height of the wave. The larger the amplitude, the bigger the movements that make the wave. The wavelength is the distance between one wave crest and the next. The frequency is the number of crests that pass a point every second. Frequency is measured in hertz (Hz).

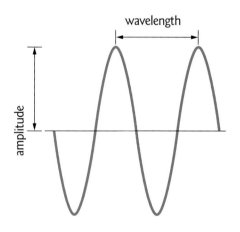

How amplitude, wavelength and frequency are measured.

Light and sound as waves

Light can be thought of as waves of **electromagnetic radiation** (see page 20). Sound travels as waves too. Sound is caused by vibrations. Imagine 'twanging' a plastic ruler on the edge of a table. The ruler vibrates up and down very quickly. These vibrations make the particles of air next to the ruler vibrate backwards and forwards. These movements are passed on to the next particles, and so waves of vibrations spread. This is how sound travels through the air. As a sound wave passes a point in the air, the particles of the air are pulled apart and then squashed together.

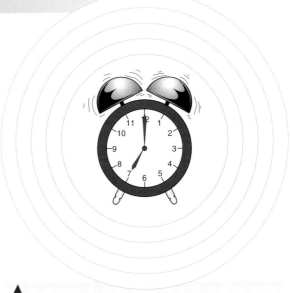

▲ Sound waves spread in all directions from a source of sound.

Wave shapes

Although you can't see a sound wave, you can draw it as a line graph. The line looks like a wave on water. It shows how the particles in the air are squeezed and stretched as time passes by. A very pure sound, such as the sound from a tuning fork, has a smooth wave shape like the one in the diagram opposite. A violin makes a much more jagged sound wave, but the same shape is repeated again and again. Noise, such as the sound of traffic, has a random wave shape. Wave shapes can be seen by connecting a microphone to an **oscilloscope**.

▼ Sections of sound waves taken from a rock music CD. They are a mixture of sounds from different musical instruments.

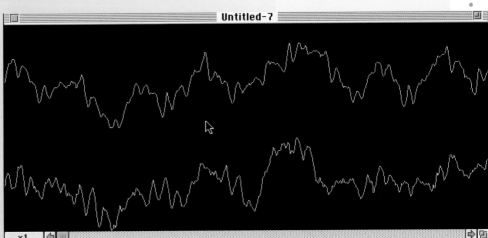

Travelling sound

Any object that vibrates causes sound, from leaves moving in the wind to the cone of a **loudspeaker**. Sound travels through solids and liquids as well as the air, but, unlike light, it cannot travel through a vacuum because it can only travel by using the particles in a substance.

Echoes

Just as waves in water bounce when they hit a surface like a harbour wall, so sound waves bounce (are reflected) when they hit surfaces. Smooth, hard surfaces, such as bare walls and cliffs, are best at reflecting sound. If you hear a sound directly from its source and then it reflects off a surface so that you hear it again, you are hearing an echo. Not all surfaces are good at reflecting sound. Soft surfaces, such as curtains, absorb sound instead. This is why an empty room, with bare walls, has echoes, but a furnished room does not.

▲ Echoes are important in a concert hall, so that the music sounds rich but without much echo. The special mushroom-shaped roof in this hall bounces sound back down to the audience.

Searching with sound

When your voice echoes, you know a surface nearby has reflected the sound. Some animals, such as dolphins and bats, create sound and use the echoes to detect objects around them. This is called **echolocation**. The time the sound takes to return indicates the object's distance. Many ships have a device which uses echolocation to detect underwater objects and to measure the water's depth.

Sound waves are also used to find underground objects, such as rocks that might contain oil and gas. Large machines fire strong sound waves into the ground, and the reflections are detected by sensitive microphones. Sound reflects off the boundaries between different types of rock. Expert **geophysicists** examine the results and try to work out the structure of the rocks.

The speed of sound

Sound travels very fast, but nothing like as fast as light. You can check this during a thunderstorm – you will see the flash of lightning well before you hear the thundering noise created by the lightning. In the air at sea-level, sound travels at about 340 metres per second. Higher in the atmosphere it slows down because the air is much thinner. At 10,000 metres upwards, just above the summit of Mount Everest, sound travels at 300 metres per second.

Sound travels much faster in liquids and solids than it does in the air. The speed of sound in water is about 1500 metres per second, and in glass it is 5000 metres per second.

Some objects, such as jet aircraft and bullets, can travel faster than the speed of sound. Their speed is normally measured by Mach numbers. Mach 1 is equal to one times the speed of sound; Mach 2 is equal to twice the speed of sound, and so on. Faster speeds than the speed of sound are called supersonic speeds. An object travelling at a supersonic speed creates a shock wave of sound, which is heard as a loud bang called a sonic boom.

▼

Humpback whales are famous for the various groaning, chirping and crying noises they make to communicate. The sound travels well underwater and can be heard more than 150 kilometres away.

Hearing sound

Our ears are for detecting sound. Just inside your ear is a thin membrane called the eardrum. When a sound wave enters your ear, the changing pressure makes the eardrum vibrate in and out. In turn this makes tiny bones attached to the eardrum vibrate, too. The bones pass the vibrations to the inner ear, where the vibrations make tiny hairs move. These movements are detected by **nerve cells** which pass messages to your brain.

Loudness and pitch

The larger the amplitude of a sound, the louder it sounds. The vibrations that cause most of the sounds we hear, such as people talking, are too small to see. But some vibrations, such as very loud drum beats, are so large you can actually feel the sound waves as well as hear them.

The higher the frequency of a sound, the more high-pitched it sounds. You can probably hear sounds with frequencies between about 20 hertz to 20,000 hertz (20 kilohertz). But when you grow older, the range of frequencies you can hear will probably be reduced. Sound with an extremely high frequency is called ultrasound. We cannot detect it but it has many applications in science, engineering and medicine.
For example, an ultrasound scanner is used to search for potentially dangerous cracks in the metal of aircraft.

Animal hearing

The range of frequencies that animals can detect is different to the range that humans can hear. For example, cats and dogs can hear sound with higher frequencies, and elephants can hear sound with lower frequencies. Bats can hear frequencies as high as 120,000 hertz, which allows them to detect the echoes of the high-pitched squeaks they use for echolocation. Many animals also have large ear flaps that allow them to hear very quiet sounds.

Dangers of sound

The loudness of a sound is measured in units called decibels (dB). On the decibel scale, an increase of 10 decibels represents a ten-times increase in loudness. This means that a sound measuring 50 decibels is ten times as loud as a sound measuring 40 decibels, and a sound of 60 decibels is 100 times as loud. Rustling leaves measure about 20 decibels; when you speak normally to your friends your speech measures about 50 decibels; and loud music measures about 100 decibels.

Sounds with a loudness of more than 120 decibels can damage your ears because the vibrations of the air are so great. Sounds this loud are caused by explosions or rockets taking off. Quieter sounds can also damage your ears if you listen to them for long enough.

Sounds can be annoying as well as dangerous. Most of the time, unless we live in the countryside, we can hear background noise such as cars on the road, aircraft flying past or music playing next door. This is called noise pollution. People who live near busy roads or airports may receive grants to help them to sound-proof their homes with double glazing. Noise is often the cause of disputes between neighbours.

▶ A jet airliner taking off creates a loud roar with its engines on full power. Older airliners, from the 1960s and 1970s, have extremely noisy engines. Many airports in urban areas have a ban on night take-offs.

Musical sounds

At its simplest, music consists of a series of sounds called notes, played one after the other. In music each note has a certain **frequency**, and playing certain notes after, or with, others makes a pleasing sound. Playing notes of other frequencies, or some notes after others, creates unpleasant sound. All musical instruments work by making the air vibrate in some way. There are three main families of acoustic instruments (that is, instruments which do not create sound electronically): string, wind and percussion.

SCIENCE ESSENTIALS

String instruments create sounds with vibrating strings.
Wind instruments create sound with a vibrating column of air.
Percussion instruments create sound with a vibrating object or diaphragm (a tightly stretched skin).

String instruments

Examples of string instruments are the guitar, violin and piano. A string instrument has a set of tight metal strings, each of which vibrates at its own natural frequency. The frequency depends on the length and thickness of the string, and how tight it is stretched. Shorter, thinner strings with more tension make higher notes than longer, fatter strings with less tension. To make the string vibrate, it is plucked (as in a guitar), hit with a hammer (as in a piano) or drawn with a bow (as in a violin). A piano has a separate string for each note. To change the note made by a string on a guitar or violin, the player shortens the string by pressing a finger on to it.

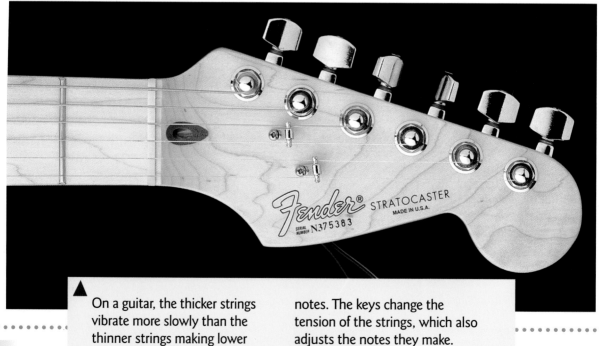

▲ On a guitar, the thicker strings vibrate more slowly than the thinner strings making lower notes. The keys change the tension of the strings, which also adjusts the notes they make.

Wind instruments

Examples of wind instruments are the trumpet, the flute and the clarinet. All wind instruments have a long tube of air inside them, and the sound they make is caused by the air in the tube vibrating. The frequency of the note that the instrument plays depends on the length of the tube. The longer the tube, the slower the air vibrates within it and the lower the notes that it can play.

The air is made to vibrate when the player blows air past a reed which then vibrates (as in a clarinet), or blows across a hole in the tube (as in a flute), or vibrates his or her lips against the end of the tube (as in a trumpet). In most wind instruments the length of the tube, and so the note the instrument plays, is changed by covering or uncovering holes in the tube, either by putting fingers over the holes or by pressing valves.

Percussion instruments

A percussion instrument makes notes when it is struck in some way, and so made to vibrate at its natural frequency. Simple percussion instruments, such as triangles and bells, only produce one note. A xylophone has many blocks of wood or metal, each of which makes its own note. Percussion instruments such as drums and gongs make noise rather than notes, but the pitch of the noise depends on their size.

► A synthesizer is an electronic musical instrument. Its complex electronic circuits can create sounds which sound almost exactly like any real instrument. They are excellent tools for composing music.

Glossary

absorption   process by which light disappears into a material

amplitude   height of a **wave** from its centre to its peak

astrophysics   the study of the physics of objects in space, such as planets, stars and galaxies

bioluminescence   light created by animals, such as glow-worms

boundary   where one material stops and another begins

chemical energy   **energy** stored in chemicals, which can be released when the chemical takes part in a **chemical reaction**

chemical reaction   when one or more chemicals divide or combine together to form new chemicals

colour spectrum   the range of colours of light, from red to violet

concave   describes a surface that dips inwards in the centre, such as the inside of a spoon

convex   describes a surface that extends outwards in the centre, such as the underside of a spoon

crest   highest point of a **wave**. The lowest point is called a trough

digital   working with numbers

dispersion   process by which different colours of light are refracted by different amounts, so splitting up **white light**

echo   sound which is heard a second time after being reflected from a surface

echolocation   method of finding objects by listening for the echoes that bounce off them

Einstein, Albert (1879–1955)   German-born Swiss physicist who developed the theory of relativity

electrical energy   **energy** carried by an electric current flowing around a circuit

electromagnetic radiation   **energy** carried from place to place by a changing magnetic and electric field

electromagnetic spectrum   family of waves, including light, radio **waves**, microwaves and X-rays

element   the simplest substance that exists, which cannot be broken into simpler substances. Hydrogen, oxygen and carbon are examples of elements

emission spectrum   mixture of different colours of light which a material emits when it is heated

energy   the ability to make something happen

enhance   to improve or make better

filter   piece of coloured glass or clear plastic that stops some colours of light but allows others to pass through

fluorescent   describes a material that gives off light when it is hit by **electromagnetic radiation**

focal point   point at which light rays from a point meet after passing through a lens or reflecting off a mirror

frequency   the number of **wave** crests that pass a point every second

galaxy   large group of stars

geophysicists   scientists who study the physics of the Earth and the Earth's structure

image   a copy in light of an object or scene

infra-red radiation   invisible rays that are similar to red light rays and carry heat **energy** from place to place

lens   specially-shaped piece of glass or clear plastic that bends light rays in an organized way

loudspeaker   device that turns a changing electric current (and electrical signal) into sound

mechanical energy   **energy** that an object has because it is moving, spinning round, or is squashed or stretched

microchip   small piece of semiconductor material (normally silicon) with a complex electronic circuit built into it

microscope   optical instrument that allows objects to be examined that are too small to be seen with the naked eye

mirror   smooth shiny surface that reflects all the light that hits it

nerve cell   animal cell that carries signals to or from the animal's brain. Nerve cells make up the animal's nervous system

nuclear energy   **energy** stored in the nucleus (centre part) of an atom, which is released when the nucleus splits up or combines with the nucleus of another atom

opaque   describes a material which does not let light pass through it

oscilloscope   instrument that shows how electrical current changes, by drawing a line on a television-like screen

periscope   optical device that turns light rays around two corners – this allows the user to get a view from above his or her head

pigments   natural chemicals which give paints, inks and animals' skins their colour

primary colours   the colours red, green and blue, which can be combined to create any colour of the **colour spectrum**

quasar   object in deep space which is thought to be a galaxy with a high-energy centre

refraction   process by which light rays change direction when they cross a boundary between two substances

reflection   process by which light or sound bounces off a surface

shadow   area of darkness created when an object blocks light

solar energy   any energy that originates from the Sun, such as sunlight

telescope   optical instrument used to examine distant objects

translucent   describes a material that scatters the light that passes through it

transmission   process by which light passes right through a material

transparent   describes a material which allows light to pass through it

ultra-violet radiation   invisible rays in sunlight, above violet in the **colour spectrum**

vacuum   a space which contains nothing, not even air

wavelength   distance between one **crest** and the next on a **wave**

waves   movements of particles (or changes in electromagnetic fields) which carry **energy** from place to place, such as waves in water, sound waves and light

white light   light made by mixing all the colours of the **colour spectrum** in equal proportions

Index